Puzzle #1

HARD

9			6					
		3			9	4		
				8	2			
4	3			1		6		
6			4					9
	2							3
		5		4		8		7
7				3				4
3		6						

Puzzle #2

HARD

			3	2				7
						4		
	6			7			8	3
	4				7	1		
			6			7	4	
					4		5	
	2	8						
	3	9				2	7	
		6		1	5			

Puzzle #3

HARD

1							9	
5								6
	8				3	7		5
	2		4				7	
				1	9	3		
					8	6		
		6	9			2		
9	3				6	1	4	
				3	7			

Puzzle #4

HARD

	8						2	
4		7		9				6
			3					
8			6	7	9		5	
	5					4		7
				2			6	
	4	3			8			5
	1			5		2		
		6						

Puzzle #5

HARD

6				3	5		8	2
3		7	9					
					4		9	
	4	1				6		
				5			3	
		6	2					
9						3	4	
	1			6	8			7
		2						8

Puzzle #6

HARD

1								7
8		5			1	4		
			5		4			
4				9		2		6
6	2		8		5			
			6		9	8		1
	3							2
		1		5		9		

Puzzle #7

HARD

8						1	6	
5	4	7						
			6					5
	3		8					9
2				7	1			
	6				3		2	
4				5				
1					2			6

Puzzle #8

HARD

				3	7			
	5		9					
3				8				9
9		2	3			6		
	4					1		
		5				3	2	
1	8				4			5
			1				4	
			2			9		

Puzzle #9

HARD

	5						3	6
		8	1		2		7	
3								
	6	1	3					
9		7	8	2			1	
						7	9	
			2	4				8
2	4			8				9

Puzzle #10

HARD

5	2				4			
				6				
7						8	2	1
1			3				8	
3	4		6			9		
	8			2	9	7		
9						6	3	
		1					5	
								8

Puzzle #11

HARD

5								
		7	2	8		5		
2	8	3						7
	4					6		
					8			2
3				6		1	5	
	6			5			3	
7								
				9	7			1

Puzzle #12

HARD

			7					2
	4	6				1	7	
			5			9		
7	6			4				
9		8						
			1	5	6			
	1					4		
	2				3			
			8				5	1

Puzzle #13

HARD

			1	2			8	
			7				5	
						6		
				6	4			7
		3				2	6	
8	7				3			4
9	5							
				8	9			
	8				1	4		6

Puzzle #14

HARD

2	4				9			
5	8	9						
							1	
3		4	7					
			1			6		3
9	5				4	2		
					6		9	5
8		3						2
						7		

Puzzle #15

HARD

6					8		5	
9		4	3			6		
5							8	
		7		4		1		
				8				
	6		9				4	
			1			3		
	4			6		7	9	
		9						1

Puzzle #16

HARD

7		8	2					
			4			5		2
					3	8		9
		6		9				
9	1			7				
			6		4			
8	2	1		6		4		
		3			8			
	4					3	6	

Puzzle #17

HARD

5				2	1	4		
	2	1		9	3		7	
		5				1		
			1		5	9		4
			9				8	5
	7			3				
	6	2				7		9
				4				6

Puzzle #18

HARD

				4	3	9		
			9				8	
	1		6					5
3	4				6		5	
		1			8			9
	7					2		4
8					2		7	
4				1				
			7					

Puzzle #19

HARD

					6			3
	7							
			7		2		8	
4								5
	6		1				4	
		3		7			1	
9			6					8
6					3		5	
		7			5	1	2	

Puzzle #20

HARD

	3	8		5				
	5				1	7		
								4
2					7	1		
9							4	5
				2			9	
6		7	1		4			9
3			6					
	9							

Puzzle #21

HARD

		5			3	8		
		3		9		4		
6	9					1	2	
	4							
			8			5		
8					2			
		4		7	8			9
					6			
	3		9		4			7

Puzzle #22

HARD

5				1			8	3
	2							
	6				3			
1			4	6		2		
							1	
	4			5				7
		7				5		8
		2		9	6			
	9	3		7				2

Puzzle #23

HARD

4						7	3	
3				9				6
	5	1						
1		4		7		9		
		6		3		2		
					8			
		7	3	1				
			5					1
		5	7				8	3

Puzzle #24

HARD

		8		4			1	
3		6			7			
2					9		3	
		5	2		3		7	
					4			2
9								
7					2	5		
								3
6	1			5				

Puzzle #25

HARD

					5	2	1	
	7			3				6
		9	1					
3	1	8			9		4	
	4	5				3		8
		6		1			9	3
1	5				2			7

Puzzle #26

HARD

			3					
	1		4			2		
5		7			2			8
7			5			8		
		2		6			7	
				8			6	
		8	9					7
9				3				4
6	5		7					

Puzzle #27

HARD

	7			8				9
			3		4		1	
8			9					
		8	1					
4								
			4		6		7	2
						2		
		1		6		5		
7	6			9		3		

Puzzle #28

HARD

				8		4		3
							6	
3	7	4						2
			2		5			
		9		1	7	6		
	8		9		2	1		
	5	7			4			6
9	4					7		

Puzzle #29

HARD

								4
				1	7	3	8	5
3	8					7		
		4	5	9		2		
			4			8		6
	3							
	2			6		9	5	
	9							
8		1	2	7				

Puzzle #30

HARD

			1		6	3		5
	4						9	
2			9			8		6
	9		2					
7				3	1			
		2	6	4				
6								
		5					2	1
						7	8	

Puzzle #31

HARD

	1			5		8	6	
7	6					1		
			3					
				6	2		9	4
6								
		9	1				3	
5					9			
	8		7	2		5		
						4	2	

Puzzle #32

HARD

				2				6
9	4							7
3		8	7				4	9
	5		6					
	1	4					3	
7		9	2		1			
			5			7		3
					6	8		
				8	3			

Puzzle #33

HARD

			9				7	
5		8			4			9
				5	8			3
	3		2					
	2			9				1
6	7				1			
				4				7
9							2	
1						4	5	

Puzzle #34

HARD

		9						
		3				1	2	4
	5		8		4		9	
	4					6		
		7		3	6			
	9	6	4		7			2
								7
				6				
				9	5	3	1	6

Puzzle #35

HARD

	3			6				
	8			4	2			1
	5						4	9
					1		9	8
	2	7	4			3		
6								
9		8					1	
		3	1			4		

Puzzle #36

HARD

	9			3				
			7		5		2	9
2					6	5	1	7
7					1	8		
	6						3	5
	8							
3		5			7			8
	2		4					6
4								

Puzzle #37

HARD

2			1				3	9
	4	3	6					
					2			
								1
	2			5			8	7
					7	6		
	8				9		5	
		7		8				4
1				3			2	

Puzzle #38

HARD

					9	5	4	2
7								
	4	2			3			
			9					
				1			8	3
	8	4	6				1	
		1						8
6					5	7		
2	5		1					

Puzzle #39

HARD

2		5			1			8
	9				8	3		
		4						
							9	
	2			9	7			1
					3	5		
8				7			1	4
					5	7		
	4		6				2	

Puzzle #40

HARD

4			8		7			6
	1					2	9	
	2		3		6		8	4
		8		2		3		
			1		3			
		7			2	1	6	
8								5

Puzzle #41

HARD

							7	5
					3	9		
	7	4		5	6			
3		1	8				5	
			1				8	3
4					9			
		9	5			7		
8				2	1		6	

Puzzle #42

HARD

4								6
		2			1			7
					5		8	
3			9					1
8				6	2		5	
9				8				
		5					7	9
		9	4		3		2	

Puzzle #43

HARD

		8						
	1							2
	7		5	1				
		6			2			1
		1	7			3		
4	9				1			
				8		2	1	4
7								6
				3	4		7	

Puzzle #44

HARD

7				9		1	5	
4			2					
		2	7	6		9		
		4						
	8						7	
	6					8	2	4
	1		9					8
					5			9
	2		6		8	5		

Puzzle #45

HARD

4		7	8	5	1			
5		3					1	
					4	3		
			3			7	8	6
				7			2	
			4	2				1
					5		6	
	6	8		9				

Puzzle #46

HARD

				4				
1					5			
3		9					8	2
9		8			3			
2		4					7	
			7					
			6		1			3
		2		3		9		
					2	5	6	

Puzzle #47

HARD

2	5			7			3	
7					9	4	2	
					5		1	6
	4			8		3		
	2		4				7	9
6		1						
			7			1		
4	6							
		5		2				

Puzzle #48

HARD

					1			8
4	1					6		
5	3		2				4	
		5				7		
				7			6	
	4				3			1
9			4					
						5		2
	6		8	3			9	

Puzzle #49

HARD

		2					4	9
	9			2				
		5	9	8			6	
4			6		3			7
							8	
	7		5			6	9	
					4		2	1
6				7				
		1	3					

Puzzle #50

HARD

6					2	1		
				9				
4		9	6					
						4	5	9
5	3							
	6		1					8
		7	4	8				
					9			5
8		3			5			7

Puzzle #51

HARD

7		4	9			3		
	3							
				4		7		1
		3		2			8	
4				3				9
						4	5	
		9		1				7
	6		8		9		2	

Puzzle #52

HARD

					1			
				2				5
			7			9	6	
4		6						
5			8					
7			6	3		8	2	
		5	2				1	
		3			8		7	
1				4	6		5	

Puzzle #53

HARD

4		6			5		8	7
5	1		6	8				
		8				3		
		4			2		3	
			3			9		
	9		4				6	
8	6			2				
3								
						5	4	

Puzzle #54

HARD

2		4			7			3
						1		
			6				8	7
7					5	2		9
9		6				4		1
3								
				4			7	
		8	1					
					9			4

Puzzle #55

HARD

								8
6	3				1	2		
1			8				5	
2	9	7			8			
			7					4
					9	5		
		6		9				
				4	2	3		
	7							

Puzzle #56

HARD

		1		2				
	5			3				7
7		2			4		6	
				8	3		7	
						5		
			9	7			2	8
								2
	3				5		1	
		7		4	6	3		

Puzzle #57

HARD

					6			9
	7			9				1
	8	2			4			
	1					4		
7			6					
6		5	3					
	3				8			6
1		7	9		3			
	4						2	

Puzzle #58

HARD

						4		3
		9			4	7		
				2	7			5
6			4		8			
		3	1					4
	1					6	9	
5	9		3	1				
4								1
2								

Puzzle #59

HARD

				4	7	2		
		1						4
7	6				5	3		
8	5		7		3		2	
			5				6	
					8			
	8				4			
				1		7		
		7	8		6			3

Puzzle #60

HARD

	3	4					7	2
6		2	3			4		
		3			1	8		9
	8	9	4			3		
1				9	6			3
	5	6		3				
				7			2	

Puzzle #61

HARD

							2	
	1		7		6		3	
				4		8		5
	8	2	9			7		
		3	6					
		7		5			1	
				1				
2					9	4		
	7		5					8

Puzzle #62

HARD

		6						9
				5				
9			1			3	7	
				9				
	8		2			7		
	5							8
7		5	4			1	6	
		1	9				3	
6				7				

Puzzle #63

HARD

1					8		5	
2			3	5				
		9		6			8	
	5					6	2	
3		2			1			4
7		3						2
	4				9			3
						5	9	

Puzzle #64

HARD

2		7					9	
	3				6			8
	1						4	
	5			9		1		
7		8			4		2	
	2					3		
			6	3				2
			7					
		1		2			7	

Puzzle #65

HARD

		5	8					
	8		1					7
				2	4	3		
	5				7	4	9	
	3					7		
8		2						
			9				1	2
6				4				
				5				

Puzzle #66

HARD

			9		2			
8			1			3		
	1							4
3		9					6	
4					8	2		
2					1	8		
	5			6			4	3
	3			7				
					5		8	

Puzzle #67

HARD

7				4				
	6				1	2		
8								
3						6		
			5		4	7		
	5	4						8
		9	4					
	3		8					6
		5		1	2	9		3

Puzzle #68

HARD

			6	5		7		
9								
5	6			1				8
						6		2
		3	2			1		9
	4					8		
7								6
	8				3		2	
1	2		7					

Puzzle #69

HARD

4	2						5	9
			5	8		4		
7								6
		6						
	7	2	6					
		8					4	3
6				1				
					8			4
			7	3			1	

Puzzle #70

HARD

			2	5	3	4		
	2					1		3
				7			9	
		4			8		7	5
1					5	8		
3				6			4	
			6					
8							6	4
					9	2		

Puzzle #71

HARD

2						1		6
	8	4				2		
			9	3				
		5	2	1			7	
	4			8				
1		6	5					9
	6		8				4	
								3
				6			5	

Puzzle #72

HARD

9			1	2				
			5		9	8		
1			6			4		
	5	1			6	7		9
							8	1
4					8	5		
	9							7
					5		6	
						3		

Puzzle #73

HARD

		8					5	
3					1			9
1					9			
	2		9		8			
		7		5		3		1
							6	
9							4	6
	5		4					7
				7				

Puzzle #74

HARD

			4	6				3
		9		1				
1	7		3			8		
6					5			4
	1			2				
2	8		7		6			
8			5				3	
					1	6		
9						5	7	

Puzzle #75

HARD

							4	
			7	8				
			6		2			9
	8	1				3		2
	3						9	
				3	1			7
1								
		5		4		2		
		2	8		6			1

Puzzle #76

HARD

							3	8
		5			8			
		4		5	1			6
			2		5			
	6	1						
9			6			7		
3						4		
			1	9			6	
				2			7	3

Puzzle #77

HARD

		2	7					4
	3					9		
		4			9			
1							6	7
5		8	2	3				
			4					
				9	5			6
	6		1	2				9
	8						4	

Puzzle #78

HARD

		1			7	8		5
6						1		
	4			5				
			6			5		9
		2	4					
								6
9	5			3				2
		8						
4						7	9	

Puzzle #79

HARD

	3				6			
6		8	7				2	
						5		
	7		4					
1	5	2				6		
			2	3			1	
				2		3		
							7	
	1		3		7			9

Puzzle #80

HARD

	1					3		4
7			6	8				
					4		9	
					6		5	
3						1		
8	6		2					
		9						
		7		5		2		9
	2		8					3

Puzzle #81

HARD

		3	9				2	
5			8			4		
	6	4						
7	1						8	
	4		7	8	2			
			3					
		5	4			6		
		7		9				
				1		8	7	

Puzzle #82

HARD

9	1				6	2		
	7		1		3	5		
4		2						
	3			5	7			1
6			3					
5			7			1	9	
			4					8
		7		8				

Puzzle #83

HARD

							6	3
		8		3		4		5
		5	1				4	
		4			6		1	
				2	7			
4						2		
7	1			8	5	3		
5		6					7	

Puzzle #84

HARD

	1			4	2			5
		8						4
	9			1			6	7
		7	8					1
9	8		2		3			
3				6				
4							1	
				3		4		
	6						2	

Puzzle #85

HARD

		6	5					
		9	8		3	4	5	
1					9			
9								
		7		6				1
3		5	2		8			7
						6		
	2				1		8	
						5		

Puzzle #86

HARD

	2			4				3
	5			2		7		
					7	1		
5								
				1			7	
		6	7			8		
	8		5					6
1	9		6					
				8		2		4

Puzzle #87

HARD

6		7						9
		8		1	9			
	5					6	2	
5							1	6
4	2			8		5		
	1		9			7		
1				2			6	
					4			7

Puzzle #88

HARD

2						4		
	3						9	
				9		8		
	9				4	2		
1	2			7	5		8	
			1					5
				5		9		2
		1		3		6		
8	5							

Puzzle #89

HARD

5							9	6
	7	4	1		8		5	
						1		8
	3	1			6			
				7				
				9				5
		8	4		3			9
		9				2	4	

Puzzle #90

HARD

		2	9	3	8	7		
8			6			9		
	7							
			5		7	4	1	
			4				6	
				6			3	5
3	9							8
5			2	1				
						1		

Puzzle #91

HARD

	3	6		4		5		
	1			6				2
			1			7	4	
		8		5				
			6				2	7
6		7						
2					9			5
			2			1	9	
		5		1				4

Puzzle #92

HARD

					6			4
7						1	9	
		1		4	3			
	9	3			1			
	2				7		3	
		9		3		2		
		2		5		4		
			4	1			8	5

Puzzle #93

HARD

		4				7		3
					2		6	
	1			9				
			7			5		1
7	3			1				
	9		2			3		
		2	5					
5	6		3			8		4
8						2		

Puzzle #94

HARD

9	6	3						
		4				3		
	8				7			
				8			3	
			5	4			7	
	5		7			1		8
5					1		4	
			6		9			7
	2	9						1

Puzzle #95

HARD

9			7			2		
	7		4	1				8
2					3			
					8		9	
7		4	3					
			2	4		3		
		3			6		5	
1				9				
8			5		4			

Puzzle #96

HARD

							7	
5	1	2				6		
		7	8	1	4			2
3		9			8		6	
1	2			6				5
8							4	
		6	5		7	8		
			2					

Puzzle #97

HARD

	8	7					1	
6			3		2			4
			5		8			
7				9			2	
						4		
	9				4			
					5	9		3
	3		4					2
9		5		6				

Puzzle #98

HARD

5	6				7			
			3	2	6			
9			1	8				4
	4					9		
		1	9					
	7				5			
	2		8				6	1
	3		6	4			2	

Puzzle #99

HARD

					8	5		
	7			5	9	2	8	
		3						9
		1	5					
	4					6		7
	9		3			8		
		7		9	5			
8						9		
		2	8	1	6			

Puzzle #100

HARD

7	6		9					8
		4	3		6			
						5		
							2	9
6				7			3	
8								
2			4	1				
	8					6		
	4	1						5

Puzzle # 1

9	7	8	6	5	4	3	2	1
2	6	3	1	7	9	4	5	8
5	1	4	3	8	2	7	9	6
4	3	9	8	1	5	6	7	2
6	5	7	4	2	3	1	8	9
8	2	1	9	6	7	5	4	3
1	9	5	2	4	6	8	3	7
7	8	2	5	3	1	9	6	4
3	4	6	7	9	8	2	1	5

Puzzle # 2

9	1	4	3	2	8	5	6	7
7	8	3	5	6	9	4	2	1
2	6	5	4	7	1	9	8	3
8	4	2	9	5	7	1	3	6
3	5	1	6	8	2	7	4	9
6	9	7	1	3	4	8	5	2
5	2	8	7	9	3	6	1	4
1	3	9	8	4	6	2	7	5
4	7	6	2	1	5	3	9	8

Puzzle # 3

1	7	2	6	5	4	8	9	3
5	9	3	7	8	2	4	1	6
6	8	4	1	9	3	7	2	5
3	2	8	4	6	5	9	7	1
7	6	5	2	1	9	3	8	4
4	1	9	3	7	8	6	5	2
8	5	6	9	4	1	2	3	7
9	3	7	5	2	6	1	4	8
2	4	1	8	3	7	5	6	9

Puzzle # 4

3	8	9	4	6	7	5	2	1
4	2	7	1	9	5	3	8	6
1	6	5	3	8	2	7	4	9
8	3	4	6	7	9	1	5	2
6	5	2	8	3	1	4	9	7
7	9	1	5	2	4	8	6	3
2	4	3	9	1	8	6	7	5
9	1	8	7	5	6	2	3	4
5	7	6	2	4	3	9	1	8

Puzzle # 5

6	9	4	7	3	5	1	8	2
3	8	7	9	1	2	4	6	5
1	2	5	6	8	4	7	9	3
2	4	1	8	7	3	6	5	9
8	7	9	4	5	6	2	3	1
5	3	6	2	9	1	8	7	4
9	5	8	1	2	7	3	4	6
4	1	3	5	6	8	9	2	7
7	6	2	3	4	9	5	1	8

Puzzle # 6

1	4	3	9	8	2	6	5	7
8	6	5	7	3	1	4	2	9
7	9	2	5	6	4	3	1	8
4	5	7	1	9	3	2	8	6
3	1	8	2	4	6	7	9	5
6	2	9	8	7	5	1	4	3
5	7	4	6	2	9	8	3	1
9	3	6	4	1	8	5	7	2
2	8	1	3	5	7	9	6	4

Puzzle # 7

8	2	9	3	4	5	1	6	7
3	1	6	2	9	7	4	5	8
5	4	7	1	6	8	3	9	2
7	8	1	6	3	9	2	4	5
6	3	5	8	2	4	7	1	9
2	9	4	5	7	1	6	8	3
9	6	8	7	1	3	5	2	4
4	7	2	9	5	6	8	3	1
1	5	3	4	8	2	9	7	6

Puzzle # 8

4	9	6	5	3	7	8	1	2
7	5	8	9	1	2	4	6	3
3	2	1	4	8	6	5	7	9
9	7	2	3	4	1	6	5	8
6	4	3	8	2	5	1	9	7
8	1	5	7	6	9	3	2	4
1	8	7	6	9	4	2	3	5
2	3	9	1	5	8	7	4	6
5	6	4	2	7	3	9	8	1

Puzzle # 9

1	5	2	7	9	4	8	3	6
6	9	8	1	3	2	4	7	5
3	7	4	6	5	8	9	2	1
4	6	1	3	7	9	5	8	2
9	3	7	8	2	5	6	1	4
8	2	5	4	6	1	7	9	3
7	1	9	2	4	6	3	5	8
5	8	6	9	1	3	2	4	7
2	4	3	5	8	7	1	6	9

Puzzle # 10

5	2	9	8	1	4	3	6	7
8	1	3	2	6	7	5	9	4
7	6	4	5	9	3	8	2	1
1	9	7	3	4	5	2	8	6
3	4	2	6	7	8	9	1	5
6	8	5	1	2	9	7	4	3
9	7	8	4	5	1	6	3	2
2	3	1	7	8	6	4	5	9
4	5	6	9	3	2	1	7	8

Puzzle # 11

5	1	6	7	4	9	8	2	3
4	9	7	2	8	3	5	1	6
2	8	3	5	1	6	4	9	7
1	4	8	3	2	5	6	7	9
6	5	9	1	7	8	3	4	2
3	7	2	9	6	4	1	5	8
9	6	1	8	5	2	7	3	4
7	2	4	6	3	1	9	8	5
8	3	5	4	9	7	2	6	1

Puzzle # 12

1	9	3	7	6	4	5	8	2
5	4	6	2	8	9	1	7	3
2	8	7	5	3	1	9	4	6
7	6	1	9	4	8	2	3	5
9	5	8	3	2	7	6	1	4
4	3	2	1	5	6	8	9	7
3	1	9	6	7	5	4	2	8
8	2	5	4	1	3	7	6	9
6	7	4	8	9	2	3	5	1

Puzzle # 13

4	3	5	1	2	6	7	8	9
6	2	9	7	4	8	1	5	3
7	1	8	3	9	5	6	4	2
2	9	1	8	6	4	5	3	7
5	4	3	9	1	7	2	6	8
8	7	6	2	5	3	9	1	4
9	5	4	6	3	2	8	7	1
1	6	7	4	8	9	3	2	5
3	8	2	5	7	1	4	9	6

Puzzle # 14

2	4	1	5	7	9	8	3	6
5	8	9	6	3	1	4	2	7
6	3	7	2	4	8	5	1	9
3	1	4	7	6	2	9	5	8
7	2	8	1	9	5	6	4	3
9	5	6	3	8	4	2	7	1
4	7	2	8	1	6	3	9	5
8	9	3	4	5	7	1	6	2
1	6	5	9	2	3	7	8	4

Puzzle # 15

6	1	2	7	9	8	4	5	3
9	8	4	3	2	5	6	1	7
5	7	3	6	1	4	2	8	9
2	9	7	5	4	6	1	3	8
4	3	5	2	8	1	9	7	6
1	6	8	9	3	7	5	4	2
8	5	6	1	7	9	3	2	4
3	4	1	8	6	2	7	9	5
7	2	9	4	5	3	8	6	1

Puzzle # 16

7	5	8	2	1	9	6	4	3
1	3	9	4	8	6	5	7	2
4	6	2	7	5	3	8	1	9
3	8	6	1	9	2	7	5	4
9	1	4	8	7	5	2	3	6
2	7	5	6	3	4	9	8	1
8	2	1	3	6	7	4	9	5
6	9	3	5	4	8	1	2	7
5	4	7	9	2	1	3	6	8

Puzzle # 17

5	8	7	6	2	1	4	9	3
6	2	1	4	9	3	5	7	8
9	4	3	8	5	7	6	1	2
2	9	5	3	8	4	1	6	7
8	3	6	1	7	5	9	2	4
7	1	4	9	6	2	3	8	5
4	7	9	2	3	6	8	5	1
3	6	2	5	1	8	7	4	9
1	5	8	7	4	9	2	3	6

Puzzle # 18

7	8	6	5	4	3	9	1	2
5	3	4	9	2	1	6	8	7
9	1	2	6	8	7	3	4	5
3	4	9	2	7	6	1	5	8
2	5	1	4	3	8	7	6	9
6	7	8	1	9	5	2	3	4
8	9	5	3	6	2	4	7	1
4	6	7	8	1	9	5	2	3
1	2	3	7	5	4	8	9	6

Puzzle # 19

1	2	8	4	5	6	7	9	3
5	7	9	3	8	1	2	6	4
3	4	6	7	9	2	5	8	1
4	9	1	2	6	8	3	7	5
7	6	5	1	3	9	8	4	2
2	8	3	5	7	4	6	1	9
9	5	2	6	1	7	4	3	8
6	1	4	8	2	3	9	5	7
8	3	7	9	4	5	1	2	6

Puzzle # 20

7	3	8	4	5	2	9	1	6
4	5	9	3	6	1	7	8	2
1	2	6	7	8	9	3	5	4
2	6	5	9	4	7	1	3	8
9	7	3	8	1	6	2	4	5
8	1	4	5	2	3	6	9	7
6	8	7	1	3	4	5	2	9
3	4	2	6	9	5	8	7	1
5	9	1	2	7	8	4	6	3

Puzzle # 21

4	1	5	7	2	3	8	9	6
2	8	3	6	9	1	4	7	5
6	9	7	4	8	5	1	2	3
3	4	2	5	6	7	9	1	8
7	6	1	8	4	9	5	3	2
8	5	9	1	3	2	7	6	4
1	2	4	3	7	8	6	5	9
9	7	8	2	5	6	3	4	1
5	3	6	9	1	4	2	8	7

Puzzle # 22

5	7	4	6	1	2	9	8	3
3	2	8	7	4	9	1	5	6
9	6	1	5	8	3	7	2	4
1	8	5	4	6	7	2	3	9
7	3	6	9	2	8	4	1	5
2	4	9	3	5	1	8	6	7
6	1	7	2	3	4	5	9	8
4	5	2	8	9	6	3	7	1
8	9	3	1	7	5	6	4	2

Puzzle # 23

4	6	9	8	5	1	7	3	2
3	2	8	4	9	7	1	5	6
7	5	1	6	2	3	8	9	4
1	3	4	2	7	5	9	6	8
5	8	6	9	3	4	2	1	7
9	7	2	1	6	8	3	4	5
8	4	7	3	1	6	5	2	9
6	9	3	5	8	2	4	7	1
2	1	5	7	4	9	6	8	3

Puzzle # 24

5	9	8	3	4	6	2	1	7
3	4	6	1	2	7	8	9	5
2	7	1	5	8	9	4	3	6
8	6	5	2	9	3	1	7	4
1	3	7	8	6	4	9	5	2
9	2	4	7	1	5	3	6	8
7	8	9	6	3	2	5	4	1
4	5	2	9	7	1	6	8	3
6	1	3	4	5	8	7	2	9

Puzzle # 25

6	3	4	8	7	5	2	1	9
2	7	1	9	3	4	5	8	6
5	8	9	1	2	6	7	3	4
3	1	8	7	5	9	6	4	2
9	4	5	2	6	1	3	7	8
7	6	2	3	4	8	9	5	1
8	2	6	5	1	7	4	9	3
1	5	3	4	9	2	8	6	7
4	9	7	6	8	3	1	2	5

Puzzle # 26

2	9	4	3	1	8	7	5	6
8	1	6	4	7	5	2	9	3
5	3	7	6	9	2	1	4	8
7	6	9	5	4	3	8	1	2
3	8	2	1	6	9	4	7	5
1	4	5	2	8	7	3	6	9
4	2	8	9	5	1	6	3	7
9	7	1	8	3	6	5	2	4
6	5	3	7	2	4	9	8	1

Puzzle # 27

1	7	3	6	8	2	4	5	9
2	9	6	3	5	4	7	1	8
8	5	4	9	1	7	6	2	3
6	2	8	1	7	5	9	3	4
4	3	7	8	2	9	1	6	5
5	1	9	4	3	6	8	7	2
3	8	5	7	4	1	2	9	6
9	4	1	2	6	3	5	8	7
7	6	2	5	9	8	3	4	1

Puzzle # 28

2	9	6	7	8	1	4	5	3
8	1	5	4	2	3	9	6	7
3	7	4	5	6	9	8	1	2
7	3	1	6	4	8	5	2	9
4	6	8	2	9	5	3	7	1
5	2	9	3	1	7	6	8	4
6	8	3	9	7	2	1	4	5
1	5	7	8	3	4	2	9	6
9	4	2	1	5	6	7	3	8

Puzzle # 29

7	1	9	3	8	5	6	2	4
2	4	6	9	1	7	3	8	5
3	8	5	6	4	2	7	1	9
1	6	4	5	9	8	2	3	7
5	7	2	4	3	1	8	9	6
9	3	8	7	2	6	5	4	1
4	2	7	1	6	3	9	5	8
6	9	3	8	5	4	1	7	2
8	5	1	2	7	9	4	6	3

Puzzle # 30

9	7	8	1	2	6	3	4	5
1	4	6	3	8	5	2	9	7
2	5	3	9	7	4	8	1	6
8	9	1	2	5	7	4	6	3
7	6	4	8	3	1	9	5	2
5	3	2	6	4	9	1	7	8
6	2	7	4	1	8	5	3	9
4	8	5	7	9	3	6	2	1
3	1	9	5	6	2	7	8	4

Puzzle # 31

9	1	2	4	5	7	8	6	3
7	6	3	2	9	8	1	4	5
8	5	4	3	1	6	9	7	2
1	3	5	8	6	2	7	9	4
6	7	8	9	3	4	2	5	1
2	4	9	1	7	5	6	3	8
5	2	1	6	4	9	3	8	7
4	8	6	7	2	3	5	1	9
3	9	7	5	8	1	4	2	6

Puzzle # 32

1	7	5	9	2	4	3	8	6
9	4	6	3	1	8	5	2	7
3	2	8	7	6	5	1	4	9
8	5	2	6	3	9	4	7	1
6	1	4	8	5	7	9	3	2
7	3	9	2	4	1	6	5	8
4	8	1	5	9	2	7	6	3
2	9	3	4	7	6	8	1	5
5	6	7	1	8	3	2	9	4

Puzzle # 33

3	1	4	9	6	2	8	7	5
5	6	8	7	3	4	2	1	9
7	9	2	1	5	8	6	4	3
8	3	1	2	7	5	9	6	4
4	2	5	6	9	3	7	8	1
6	7	9	4	8	1	5	3	2
2	5	3	8	4	6	1	9	7
9	4	6	5	1	7	3	2	8
1	8	7	3	2	9	4	5	6

Puzzle # 34

4	1	9	2	7	3	8	6	5
8	7	3	6	5	9	1	2	4
6	5	2	8	1	4	7	9	3
3	4	8	5	2	1	6	7	9
5	2	7	9	3	6	4	8	1
1	9	6	4	8	7	5	3	2
9	6	1	3	4	8	2	5	7
7	3	5	1	6	2	9	4	8
2	8	4	7	9	5	3	1	6

Puzzle # 35

4	3	1	9	6	8	5	2	7
7	8	9	5	4	2	6	3	1
2	5	6	3	1	7	8	4	9
8	9	5	6	2	3	1	7	4
3	6	4	7	5	1	2	9	8
1	2	7	4	8	9	3	6	5
6	1	2	8	7	4	9	5	3
9	4	8	2	3	5	7	1	6
5	7	3	1	9	6	4	8	2

Puzzle # 36

5	9	7	1	3	2	6	8	4
6	4	1	7	8	5	3	2	9
2	3	8	9	4	6	5	1	7
7	5	4	3	6	1	8	9	2
1	6	2	8	7	9	4	3	5
9	8	3	5	2	4	7	6	1
3	1	5	6	9	7	2	4	8
8	2	9	4	5	3	1	7	6
4	7	6	2	1	8	9	5	3

Puzzle # 37

2	6	8	1	7	5	4	3	9
7	4	3	6	9	8	2	1	5
5	1	9	3	4	2	8	7	6
6	7	4	8	2	3	5	9	1
9	2	1	4	5	6	3	8	7
8	3	5	9	1	7	6	4	2
4	8	2	7	6	9	1	5	3
3	5	7	2	8	1	9	6	4
1	9	6	5	3	4	7	2	8

Puzzle # 38

3	1	6	7	8	9	5	4	2
7	9	5	4	2	1	8	3	6
8	4	2	5	6	3	1	7	9
1	2	3	9	5	8	4	6	7
5	6	7	2	1	4	9	8	3
9	8	4	6	3	7	2	1	5
4	7	1	3	9	2	6	5	8
6	3	9	8	4	5	7	2	1
2	5	8	1	7	6	3	9	4

Puzzle # 39

2	3	5	9	4	1	6	7	8
6	9	1	7	5	8	3	4	2
7	8	4	2	3	6	1	5	9
1	7	8	5	6	4	2	9	3
5	2	3	8	9	7	4	6	1
4	6	9	1	2	3	5	8	7
8	5	6	3	7	2	9	1	4
9	1	2	4	8	5	7	3	6
3	4	7	6	1	9	8	2	5

Puzzle # 40

4	9	2	8	1	7	5	3	6
7	1	3	6	4	5	2	9	8
5	8	6	2	3	9	4	7	1
1	2	9	3	5	6	7	8	4
3	5	4	9	7	8	6	1	2
6	7	8	4	2	1	3	5	9
2	6	5	1	9	3	8	4	7
9	4	7	5	8	2	1	6	3
8	3	1	7	6	4	9	2	5

Puzzle # 41

6	1	3	4	9	8	2	7	5
5	2	8	7	1	3	9	4	6
9	7	4	2	5	6	8	3	1
3	6	1	8	7	2	4	5	9
7	9	2	1	4	5	6	8	3
4	8	5	6	3	9	1	2	7
2	3	9	5	6	4	7	1	8
8	5	7	9	2	1	3	6	4
1	4	6	3	8	7	5	9	2

Puzzle # 42

4	3	8	7	2	9	5	1	6
5	6	2	8	3	1	4	9	7
7	9	1	6	4	5	2	8	3
3	2	4	9	5	7	8	6	1
8	1	7	3	6	2	9	5	4
9	5	6	1	8	4	7	3	2
6	4	5	2	1	8	3	7	9
2	7	3	5	9	6	1	4	8
1	8	9	4	7	3	6	2	5

Puzzle # 43

5	4	8	2	6	3	1	9	7
6	1	3	4	7	9	5	8	2
2	7	9	5	1	8	4	6	3
3	5	6	8	9	2	7	4	1
8	2	1	7	4	6	3	5	9
4	9	7	3	5	1	6	2	8
9	3	5	6	8	7	2	1	4
7	8	4	1	2	5	9	3	6
1	6	2	9	3	4	8	7	5

Puzzle # 44

7	3	6	8	9	4	1	5	2
4	9	1	2	5	3	7	8	6
8	5	2	7	6	1	9	4	3
2	7	4	5	8	6	3	9	1
1	8	3	4	2	9	6	7	5
9	6	5	3	1	7	8	2	4
5	1	7	9	3	2	4	6	8
6	4	8	1	7	5	2	3	9
3	2	9	6	4	8	5	1	7

Puzzle # 45

4	2	7	8	5	1	6	9	3
8	1	6	9	3	2	4	7	5
5	9	3	6	4	7	2	1	8
7	8	1	2	6	4	3	5	9
2	5	4	3	1	9	7	8	6
6	3	9	5	7	8	1	2	4
9	7	5	4	2	6	8	3	1
3	4	2	1	8	5	9	6	7
1	6	8	7	9	3	5	4	2

Puzzle # 46

8	2	6	3	4	9	7	1	5
1	4	7	8	2	5	3	9	6
3	5	9	1	6	7	4	8	2
9	7	8	2	1	3	6	5	4
2	3	4	9	5	6	1	7	8
5	6	1	7	8	4	2	3	9
4	9	5	6	7	1	8	2	3
6	1	2	5	3	8	9	4	7
7	8	3	4	9	2	5	6	1

Puzzle # 47

2	5	4	6	7	1	9	3	8
7	1	6	8	3	9	4	2	5
8	3	9	2	4	5	7	1	6
9	4	7	5	8	2	3	6	1
5	2	3	4	1	6	8	7	9
6	8	1	3	9	7	5	4	2
3	9	2	7	6	8	1	5	4
4	6	8	1	5	3	2	9	7
1	7	5	9	2	4	6	8	3

Puzzle # 48

6	7	9	3	4	1	2	5	8
4	1	2	5	8	9	6	7	3
5	3	8	2	6	7	1	4	9
1	9	5	6	2	8	7	3	4
8	2	3	1	7	4	9	6	5
7	4	6	9	5	3	8	2	1
9	5	7	4	1	2	3	8	6
3	8	4	7	9	6	5	1	2
2	6	1	8	3	5	4	9	7

Puzzle # 49

8	1	2	7	3	6	5	4	9
7	9	6	4	2	5	1	3	8
3	4	5	9	8	1	7	6	2
4	5	8	6	9	3	2	1	7
1	6	9	2	4	7	3	8	5
2	7	3	5	1	8	6	9	4
5	3	7	8	6	4	9	2	1
6	2	4	1	7	9	8	5	3
9	8	1	3	5	2	4	7	6

Puzzle # 50

6	8	5	7	4	2	1	9	3
3	7	2	5	9	1	8	6	4
4	1	9	6	3	8	5	7	2
7	2	1	8	6	3	4	5	9
5	3	8	9	2	4	7	1	6
9	6	4	1	5	7	3	2	8
2	5	7	4	8	6	9	3	1
1	4	6	3	7	9	2	8	5
8	9	3	2	1	5	6	4	7

Puzzle # 51

7	1	4	9	8	2	3	6	5
9	3	5	7	6	1	8	4	2
6	2	8	5	4	3	7	9	1
5	9	3	4	2	7	1	8	6
4	8	6	1	3	5	2	7	9
1	7	2	6	9	8	4	5	3
8	5	9	2	1	4	6	3	7
2	4	7	3	5	6	9	1	8
3	6	1	8	7	9	5	2	4

Puzzle # 52

9	7	8	5	6	1	3	4	2
3	6	1	4	2	9	7	8	5
2	5	4	7	8	3	9	6	1
4	8	6	9	1	2	5	3	7
5	3	2	8	7	4	1	9	6
7	1	9	6	3	5	8	2	4
8	4	5	2	9	7	6	1	3
6	2	3	1	5	8	4	7	9
1	9	7	3	4	6	2	5	8

Puzzle # 53

4	3	6	2	9	5	1	8	7
5	1	7	6	8	3	4	9	2
9	2	8	1	4	7	3	5	6
7	5	4	9	6	2	8	3	1
6	8	2	3	5	1	9	7	4
1	9	3	4	7	8	2	6	5
8	6	9	5	2	4	7	1	3
3	4	5	7	1	9	6	2	8
2	7	1	8	3	6	5	4	9

Puzzle # 54

2	5	4	8	1	7	6	9	3
8	6	7	9	5	3	1	4	2
1	9	3	6	2	4	5	8	7
7	8	1	4	6	5	2	3	9
9	2	6	7	3	8	4	5	1
3	4	5	2	9	1	7	6	8
5	1	9	3	4	2	8	7	6
4	3	8	1	7	6	9	2	5
6	7	2	5	8	9	3	1	4

Puzzle # 55

7	5	4	9	2	3	1	6	8
6	3	8	5	7	1	2	4	9
1	2	9	8	6	4	7	5	3
2	9	7	4	5	8	6	3	1
8	1	5	7	3	6	9	2	4
4	6	3	2	1	9	5	8	7
5	4	6	3	9	7	8	1	2
9	8	1	6	4	2	3	7	5
3	7	2	1	8	5	4	9	6

Puzzle # 56

3	8	1	6	2	7	9	5	4
4	5	6	1	3	9	2	8	7
7	9	2	8	5	4	1	6	3
6	2	9	5	8	3	4	7	1
1	7	8	4	6	2	5	3	9
5	4	3	9	7	1	6	2	8
9	6	5	3	1	8	7	4	2
2	3	4	7	9	5	8	1	6
8	1	7	2	4	6	3	9	5

Puzzle # 57

4	5	1	7	3	6	2	8	9
3	7	6	8	9	2	5	4	1
9	8	2	5	1	4	6	3	7
8	1	3	2	7	9	4	6	5
7	9	4	6	8	5	3	1	2
6	2	5	3	4	1	7	9	8
2	3	9	4	5	8	1	7	6
1	6	7	9	2	3	8	5	4
5	4	8	1	6	7	9	2	3

Puzzle # 58

7	6	2	9	5	1	4	8	3
1	5	9	8	3	4	7	2	6
3	4	8	6	2	7	9	1	5
6	2	5	4	9	8	1	3	7
9	7	3	1	6	2	8	5	4
8	1	4	5	7	3	6	9	2
5	9	7	3	1	6	2	4	8
4	3	6	2	8	9	5	7	1
2	8	1	7	4	5	3	6	9

Puzzle # 59

9	3	8	1	4	7	2	5	6
5	2	1	6	3	9	8	7	4
7	6	4	2	8	5	3	1	9
8	5	9	7	6	3	4	2	1
4	7	3	5	2	1	9	6	8
6	1	2	4	9	8	5	3	7
1	8	5	3	7	4	6	9	2
3	4	6	9	1	2	7	8	5
2	9	7	8	5	6	1	4	3

Puzzle # 60

9	3	4	6	5	8	1	7	2
8	1	5	2	4	7	9	3	6
6	7	2	3	1	9	4	8	5
5	4	3	7	2	1	8	6	9
7	6	1	9	8	3	2	5	4
2	8	9	4	6	5	3	1	7
1	2	7	8	9	6	5	4	3
4	5	6	1	3	2	7	9	8
3	9	8	5	7	4	6	2	1

Puzzle # 61

6	3	4	8	9	5	1	2	7
8	1	5	7	2	6	9	3	4
7	2	9	1	4	3	8	6	5
5	8	2	9	3	1	7	4	6
1	4	3	6	7	2	5	8	9
9	6	7	4	5	8	3	1	2
4	9	8	2	1	7	6	5	3
2	5	6	3	8	9	4	7	1
3	7	1	5	6	4	2	9	8

Puzzle # 62

5	1	6	3	2	7	4	8	9
4	7	3	8	5	9	6	2	1
9	2	8	1	4	6	3	7	5
1	6	4	7	9	8	2	5	3
3	8	9	2	1	5	7	4	6
2	5	7	6	3	4	9	1	8
7	9	5	4	8	3	1	6	2
8	4	1	9	6	2	5	3	7
6	3	2	5	7	1	8	9	4

Puzzle # 63

9	7	5	1	4	6	2	3	8
1	3	4	9	2	8	7	5	6
2	8	6	3	5	7	4	1	9
4	1	9	7	6	2	3	8	5
8	5	7	4	9	3	6	2	1
3	6	2	5	8	1	9	7	4
7	9	3	6	1	5	8	4	2
5	4	8	2	7	9	1	6	3
6	2	1	8	3	4	5	9	7

Puzzle # 64

2	8	7	4	5	3	6	9	1
9	3	4	1	7	6	2	5	8
5	1	6	9	8	2	7	4	3
4	5	3	2	9	8	1	6	7
7	6	8	3	1	4	5	2	9
1	2	9	5	6	7	3	8	4
8	7	5	6	3	9	4	1	2
6	9	2	7	4	1	8	3	5
3	4	1	8	2	5	9	7	6

Puzzle # 65

7	2	5	8	6	3	9	4	1
3	8	4	1	9	5	2	6	7
9	6	1	7	2	4	3	8	5
1	5	6	2	8	7	4	9	3
4	3	9	5	1	6	7	2	8
8	7	2	4	3	9	1	5	6
5	4	3	9	7	8	6	1	2
6	1	8	3	4	2	5	7	9
2	9	7	6	5	1	8	3	4

Puzzle # 66

6	4	3	9	5	2	1	7	8
8	9	7	1	4	6	3	2	5
5	1	2	7	8	3	6	9	4
3	8	9	5	2	7	4	6	1
4	7	1	6	3	8	2	5	9
2	6	5	4	9	1	8	3	7
1	5	8	2	6	9	7	4	3
9	3	6	8	7	4	5	1	2
7	2	4	3	1	5	9	8	6

Puzzle # 67

7	9	2	3	4	8	5	6	1
5	6	3	9	7	1	2	8	4
8	4	1	2	5	6	3	9	7
3	7	8	1	2	9	6	4	5
1	2	6	5	8	4	7	3	9
9	5	4	7	6	3	1	2	8
6	1	9	4	3	7	8	5	2
2	3	7	8	9	5	4	1	6
4	8	5	6	1	2	9	7	3

Puzzle # 68

3	1	8	6	5	2	7	9	4
9	7	2	3	8	4	5	6	1
5	6	4	9	1	7	2	3	8
8	9	7	4	3	1	6	5	2
6	5	3	2	7	8	1	4	9
2	4	1	5	9	6	8	7	3
7	3	9	8	2	5	4	1	6
4	8	5	1	6	3	9	2	7
1	2	6	7	4	9	3	8	5

Puzzle # 69

4	2	1	3	6	7	8	5	9
9	6	3	5	8	1	4	2	7
7	8	5	4	2	9	1	3	6
3	4	6	8	9	5	2	7	1
1	7	2	6	4	3	5	9	8
5	9	8	1	7	2	6	4	3
6	3	9	2	1	4	7	8	5
2	1	7	9	5	8	3	6	4
8	5	4	7	3	6	9	1	2

Puzzle # 70

9	1	6	2	5	3	4	8	7
7	2	8	9	4	6	1	5	3
4	5	3	8	7	1	6	9	2
2	6	4	1	9	8	3	7	5
1	7	9	4	3	5	8	2	6
3	8	5	7	6	2	9	4	1
5	3	2	6	8	4	7	1	9
8	9	1	3	2	7	5	6	4
6	4	7	5	1	9	2	3	8

Puzzle # 71

2	5	9	7	4	8	1	3	6
3	8	4	6	5	1	2	9	7
6	1	7	9	3	2	4	8	5
8	9	5	2	1	6	3	7	4
7	4	2	3	8	9	5	6	1
1	3	6	5	7	4	8	2	9
5	6	1	8	9	3	7	4	2
9	7	8	4	2	5	6	1	3
4	2	3	1	6	7	9	5	8

Puzzle # 72

9	8	3	1	2	4	6	7	5
2	4	6	5	7	9	8	1	3
1	7	5	6	8	3	4	9	2
8	5	1	2	3	6	7	4	9
6	3	9	4	5	7	2	8	1
4	2	7	9	1	8	5	3	6
3	9	4	8	6	2	1	5	7
7	1	2	3	4	5	9	6	8
5	6	8	7	9	1	3	2	4

Puzzle # 73

7	9	8	3	6	4	1	5	2
3	4	5	7	2	1	6	8	9
1	6	2	5	8	9	7	3	4
6	2	1	9	3	8	4	7	5
4	8	7	2	5	6	3	9	1
5	3	9	1	4	7	2	6	8
9	7	3	8	1	2	5	4	6
2	5	6	4	9	3	8	1	7
8	1	4	6	7	5	9	2	3

Puzzle # 74

5	2	8	4	6	7	1	9	3
4	3	9	2	1	8	7	5	6
1	7	6	3	5	9	8	4	2
6	9	7	1	3	5	2	8	4
3	1	5	8	2	4	9	6	7
2	8	4	7	9	6	3	1	5
8	6	1	5	7	2	4	3	9
7	5	3	9	4	1	6	2	8
9	4	2	6	8	3	5	7	1

Puzzle # 75

6	2	7	9	1	3	8	4	5
5	1	9	7	8	4	6	2	3
8	4	3	6	5	2	7	1	9
7	8	1	4	6	9	3	5	2
2	3	4	5	7	8	1	9	6
9	5	6	2	3	1	4	8	7
1	6	8	3	2	5	9	7	4
3	9	5	1	4	7	2	6	8
4	7	2	8	9	6	5	3	1

Puzzle # 76

6	1	9	7	4	2	5	3	8
2	3	5	9	6	8	1	4	7
8	7	4	3	5	1	2	9	6
7	4	3	2	1	5	6	8	9
5	6	1	8	7	9	3	2	4
9	8	2	6	3	4	7	5	1
3	9	6	5	8	7	4	1	2
4	2	7	1	9	3	8	6	5
1	5	8	4	2	6	9	7	3

Puzzle # 77

8	9	2	7	6	1	5	3	4
6	3	1	5	4	2	9	7	8
7	5	4	3	8	9	6	1	2
1	4	3	9	5	8	2	6	7
5	7	8	2	3	6	4	9	1
9	2	6	4	1	7	8	5	3
4	1	7	8	9	5	3	2	6
3	6	5	1	2	4	7	8	9
2	8	9	6	7	3	1	4	5

Puzzle # 78

2	9	1	3	4	7	8	6	5
6	7	5	9	8	2	1	3	4
8	4	3	1	5	6	9	2	7
7	1	4	6	2	3	5	8	9
5	6	2	4	9	8	3	7	1
3	8	9	5	7	1	2	4	6
9	5	7	8	3	4	6	1	2
1	2	8	7	6	9	4	5	3
4	3	6	2	1	5	7	9	8

Puzzle # 79

2	3	5	1	4	6	7	9	8
6	9	8	7	5	3	1	2	4
7	4	1	8	9	2	5	3	6
3	7	9	4	6	1	8	5	2
1	5	2	9	7	8	6	4	3
4	8	6	2	3	5	9	1	7
9	6	7	5	2	4	3	8	1
8	2	3	6	1	9	4	7	5
5	1	4	3	8	7	2	6	9

Puzzle # 80

9	1	6	7	2	5	3	8	4
7	4	3	6	8	9	5	2	1
2	5	8	3	1	4	7	9	6
1	7	2	9	3	6	4	5	8
3	9	4	5	7	8	1	6	2
8	6	5	2	4	1	9	3	7
4	3	9	1	6	2	8	7	5
6	8	7	4	5	3	2	1	9
5	2	1	8	9	7	6	4	3

Puzzle # 81

8	7	3	9	5	4	1	2	6
5	9	1	8	2	6	4	3	7
2	6	4	1	3	7	9	5	8
7	1	2	5	6	9	3	8	4
3	4	9	7	8	2	5	6	1
6	5	8	3	4	1	7	9	2
9	2	5	4	7	8	6	1	3
1	8	7	6	9	3	2	4	5
4	3	6	2	1	5	8	7	9

Puzzle # 82

9	1	3	5	4	6	2	8	7
8	7	6	1	2	3	5	4	9
4	5	2	9	7	8	6	1	3
7	9	1	2	6	4	8	3	5
2	3	4	8	5	7	9	6	1
6	8	5	3	9	1	4	7	2
5	4	8	7	3	2	1	9	6
3	6	9	4	1	5	7	2	8
1	2	7	6	8	9	3	5	4

Puzzle # 83

2	4	7	5	1	8	9	6	3
1	6	8	7	3	9	4	2	5
9	5	3	2	6	4	1	8	7
8	7	5	1	9	3	6	4	2
3	2	4	8	5	6	7	1	9
6	9	1	4	2	7	5	3	8
4	8	9	3	7	1	2	5	6
7	1	2	6	8	5	3	9	4
5	3	6	9	4	2	8	7	1

Puzzle # 84

7	1	6	9	4	2	8	3	5
2	3	8	6	7	5	1	9	4
5	9	4	3	1	8	2	6	7
6	2	7	8	9	4	3	5	1
9	8	1	2	5	3	7	4	6
3	4	5	7	6	1	9	8	2
4	7	3	5	2	9	6	1	8
8	5	2	1	3	6	4	7	9
1	6	9	4	8	7	5	2	3

Puzzle # 85

8	3	6	5	7	4	1	9	2
2	7	9	8	1	3	4	5	6
1	5	4	6	2	9	3	7	8
9	6	2	1	3	7	8	4	5
4	8	7	9	6	5	2	3	1
3	1	5	2	4	8	9	6	7
5	4	8	7	9	2	6	1	3
6	2	3	4	5	1	7	8	9
7	9	1	3	8	6	5	2	4

Puzzle # 86

8	2	7	1	4	5	6	9	3
6	5	1	3	2	9	7	4	8
3	4	9	8	6	7	1	5	2
5	7	2	4	9	8	3	6	1
9	3	8	2	1	6	4	7	5
4	1	6	7	5	3	8	2	9
2	8	3	5	7	4	9	1	6
1	9	4	6	3	2	5	8	7
7	6	5	9	8	1	2	3	4

Puzzle # 87

6	3	7	2	4	5	1	8	9
2	4	8	6	1	9	3	7	5
9	5	1	8	7	3	6	2	4
5	7	9	4	3	2	8	1	6
4	2	6	7	8	1	5	9	3
8	1	3	9	5	6	7	4	2
1	9	5	3	2	7	4	6	8
7	6	4	5	9	8	2	3	1
3	8	2	1	6	4	9	5	7

Puzzle # 88

2	8	9	5	1	3	4	6	7
7	3	6	2	4	8	5	9	1
4	1	5	7	9	6	8	2	3
5	9	7	3	8	4	2	1	6
1	2	4	6	7	5	3	8	9
3	6	8	1	2	9	7	4	5
6	4	3	8	5	1	9	7	2
9	7	1	4	3	2	6	5	8
8	5	2	9	6	7	1	3	4

Puzzle # 89

5	1	2	7	3	4	8	9	6
3	8	6	5	2	9	4	7	1
9	7	4	1	6	8	3	5	2
2	9	7	3	4	5	1	6	8
4	3	1	9	8	6	5	2	7
8	6	5	2	7	1	9	3	4
1	4	3	6	9	2	7	8	5
7	2	8	4	5	3	6	1	9
6	5	9	8	1	7	2	4	3

Puzzle # 90

4	1	2	9	3	8	7	5	6
8	5	3	6	7	2	9	4	1
9	7	6	1	5	4	3	8	2
6	3	8	5	2	7	4	1	9
1	2	5	4	9	3	8	6	7
7	4	9	8	6	1	2	3	5
3	9	1	7	4	6	5	2	8
5	8	4	2	1	9	6	7	3
2	6	7	3	8	5	1	9	4

Puzzle # 91

7	3	6	8	4	2	5	1	9
9	1	4	5	6	7	3	8	2
5	8	2	1	9	3	7	4	6
3	2	8	7	5	4	9	6	1
1	5	9	6	3	8	4	2	7
6	4	7	9	2	1	8	5	3
2	7	1	4	8	9	6	3	5
4	6	3	2	7	5	1	9	8
8	9	5	3	1	6	2	7	4

Puzzle # 92

9	5	8	1	7	6	3	2	4
7	3	4	2	8	5	1	9	6
2	6	1	9	4	3	5	7	8
8	9	3	5	6	1	7	4	2
4	2	5	8	9	7	6	3	1
6	1	7	3	2	4	8	5	9
5	4	9	6	3	8	2	1	7
1	8	2	7	5	9	4	6	3
3	7	6	4	1	2	9	8	5

Puzzle # 93

9	2	4	6	5	1	7	8	3
3	5	8	4	7	2	1	6	9
6	1	7	8	9	3	4	2	5
2	8	6	7	3	4	5	9	1
7	3	5	9	1	8	6	4	2
4	9	1	2	6	5	3	7	8
1	4	2	5	8	6	9	3	7
5	6	9	3	2	7	8	1	4
8	7	3	1	4	9	2	5	6

Puzzle # 94

9	6	3	4	1	5	7	8	2
7	1	4	2	6	8	3	9	5
2	8	5	9	3	7	4	1	6
6	9	7	1	8	2	5	3	4
1	3	8	5	4	6	2	7	9
4	5	2	7	9	3	1	6	8
5	7	6	8	2	1	9	4	3
3	4	1	6	5	9	8	2	7
8	2	9	3	7	4	6	5	1

Puzzle # 95

9	3	8	7	6	5	2	4	1
6	7	5	4	1	2	9	3	8
2	4	1	9	8	3	7	6	5
3	1	2	6	7	8	5	9	4
7	6	4	3	5	9	1	8	2
5	8	9	2	4	1	3	7	6
4	9	3	1	2	6	8	5	7
1	5	6	8	9	7	4	2	3
8	2	7	5	3	4	6	1	9

Puzzle # 96

4	8	3	6	5	2	9	7	1
5	1	2	3	7	9	6	8	4
6	9	7	8	1	4	5	3	2
7	6	4	1	9	5	3	2	8
3	5	9	4	2	8	1	6	7
1	2	8	7	6	3	4	9	5
8	7	5	9	3	1	2	4	6
2	3	6	5	4	7	8	1	9
9	4	1	2	8	6	7	5	3

Puzzle # 97

3	8	7	9	4	6	2	1	5
6	5	9	3	1	2	8	7	4
4	1	2	5	7	8	6	3	9
7	4	3	8	9	1	5	2	6
5	6	8	2	3	7	4	9	1
2	9	1	6	5	4	3	8	7
8	7	4	1	2	5	9	6	3
1	3	6	4	8	9	7	5	2
9	2	5	7	6	3	1	4	8

Puzzle # 98

2	1	3	5	9	8	4	7	6
5	6	8	4	1	7	2	9	3
7	9	4	3	2	6	1	8	5
9	5	7	1	8	2	6	3	4
6	4	2	7	5	3	9	1	8
3	8	1	9	6	4	7	5	2
1	7	6	2	3	5	8	4	9
4	2	5	8	7	9	3	6	1
8	3	9	6	4	1	5	2	7

Puzzle # 99

1	2	9	4	3	8	5	7	6
6	7	4	1	5	9	2	8	3
5	8	3	6	7	2	4	1	9
7	6	1	5	8	4	3	9	2
3	4	8	9	2	1	6	5	7
2	9	5	3	6	7	8	4	1
4	3	7	2	9	5	1	6	8
8	1	6	7	4	3	9	2	5
9	5	2	8	1	6	7	3	4

Puzzle # 100

7	6	3	9	5	2	4	1	8
1	5	4	3	8	6	9	7	2
9	2	8	7	4	1	5	6	3
4	7	5	6	3	8	1	2	9
6	1	2	5	7	9	8	3	4
8	3	9	1	2	4	7	5	6
2	9	6	4	1	5	3	8	7
5	8	7	2	9	3	6	4	1
3	4	1	8	6	7	2	9	5

www.ingramcontent.com/pod-product-compliance
Lightning Source LLC
Chambersburg PA
CBHW080921170526
45158CB00008B/2192
9781077210363